Welcome to Your 4th Grade Math Adventure!

Dear Student,

You are about to begin an exciting journey through 4th grade math! This workbook was created just for YOU — to help you build confidence AND develop strong math skills.

Why Practice Matters

Math is like a muscle — the more you use it, the stronger it gets! Practicing helps you:

- Master important skills like multiplication, division, and place value
- Solve real-world problems with confidence
- Prepare for quizzes, tests, and success in the next grade
- Build **focus, discipline, and problem-solving** superpowers

What You'll Learn

The Measurements and Data workbook covers:

- Solving word problems involving length, mass, volume, time & Money
- Converting units within the same measurement system.
- Understanding area and perimeter
- Understanding types of angles (acute, obtuse, right)
- Representing and interpreting data using graphs and pictographs

You Can Do This!

Some problems will be easy. Others will challenge you. That's okay! Every page you complete is a victory. Don't be afraid to make mistakes — they are part of learning. Take your time, use scratch paper, and don't forget to **ask questions** when you need help.

Our Goal

To help you become a confident problem-solver, critical thinker, and math rockstar!

4th Grade Math Workbook

Workbook Index

☑ Student Goals & Progress Checklist

Instructions: Check off each box after you complete the section. You can also write in the date you completed it!

Section	Completed	Date	Parent/Teacher Initial
Let's Learn About It!	☐	______	______
Examples & Step-by-Step Practice	☐	______	______
Pre-Test	☐	______	______
Word Problems Using Length, Mass, Volume, Type, & Money	☐	______	______
Unit Conversion	☐	______	______
Understanding Area & Perimeter	☐	______	______
Identify the Type of Angle	☐	______	______
Find the Perimeter	☐	______	______
Convert the Measurements	☐	______	______
Multi-Step Problems	☐	______	______
Representing & Interpreting Data	☐	______	______
Post Test	☐		
Certificate of Completion 🎓	☐	______	______

Let's Learn About it!

Understanding Angles: Acute, Right, and Obtuse

Step 1: Understand what an angle is

An angle is formed when two lines (or rays) meet at a point called the vertex. The amount of turn between the two lines is measured in degrees (°).

Step 2: Define each type of angle

◆ **Acute Angle**

Definition: An angle that measures less than 90°.
Example: 45°, 30°, 75°
Visual: Looks like a sharp, narrow angle.

◆ **Right Angle**

Definition: An angle that measures exactly 90°.
Example: 90°
Visual: Forms a perfect "L" shape, like the corner of a square.

◆ **Obtuse Angle**

Definition: An angle that measures more than 90° but less than 180°.
Example: 120°, 135°, 100°
Visual: Appears wide or stretched out.

Step 3: Compare the angles

Type of Angle	Degree Range	Description
Acute	Less than 90°	Sharp and narrow
Right	Exactly 90°	Perfect corner (L-shape)
Obtuse	Between 90°–180°	Wide and open

✅ Summary

- Acute = small angle (< 90°)
- Right = perfect angle (90°)
- Obtuse = large angle (> 90° but < 180°)

Conversion Table

This table helps you convert units easily. Use it as a quick reference.

Length Conversion Table

Unit	Equivalent
1 kilometer	1000 meters
1 meter	100 centimeters
1 centimeter	10 millimeters

Mass Conversion Table

Unit	Equivalent
1 kilogram	1000 grams
1 gram	1000 milligrams

Volume Conversion Table

Unit	Equivalent
1 liter	1000 milliliters
1 milliliter	0.001 liter

Time Conversion Table

Unit	Equivalent
1 hour	60 minutes
1 minute	60 seconds
1 day	24 hours

PRO TIPS

- Always label your answers (e.g., books, miles).
- Estimate first to check if your answer makes sense.
- Show your work clearly to avoid mistakes.

Understanding Area and Perimeter

Definitions

• Perimeter: The total distance around a shape.
• Area: The amount of space inside a shape.

Key Formulas

Rectangle

• Perimeter = 2 × (Length + Width)
• Area = Length × Width

Example: Length = 8 m, Width = 5 m → Perimeter = 2 × (8 + 5) = 26 m; Area = 8 × 5 = 40 m^2

Square

• Perimeter = 4 × Side
• Area = Side × Side

Example: Side = 6 cm → Perimeter = 4 × 6 = 24 cm; Area = 6 × 6 = 36 cm^2

Triangle

• Perimeter = Sum of all sides
• Area = (Base × Height) ÷ 2

Example: Base = 10 cm, Height = 5 cm → Area = (10 × 5) ÷ 2 = 25 cm^2

Parallelogram

• Perimeter = 2 × (Base + Side)
• Area = Base × Height

Example: Base = 12 m, Height = 7 m → Area = 12 × 7 = 84 m^2

Trapezoid

• Perimeter = Sum of all sides
• Area = ((Base1 + Base2) ÷ 2) × Height

Example: Base1 = 8 cm, Base2 = 6 cm, Height = 4 cm → Area = ((8 + 6) ÷ 2) × 4 = 28 cm^2

Circle

• Perimeter (Circumference) = 2 × π × Radius
• Area = π × Radius^2

Example: Radius = 7 cm → Circumference ≈ 2 × 3.14 × 7 = 43.96 cm; Area ≈ 3.14 × 7^2 = 153.86 cm^2

4th Grade Math Workbook
Measurement & Data Examples

1. Solving Word Problems (Length, Mass, Volume, Time & Money)
Example:
Liam buys 3 pencils for $0.75 each and a notebook for $2.50. How much money does he spend in total?
Steps:
1. **Identify costs:**
   - Pencil = $0.75 each
   - Notebook = $2.50
2. **Multiply pencil cost by quantity:**
   $$0.75 \times 3 = 2.25$$
3. **Add notebook cost:**
   $$2.25 + 2.50 = 4.75$$
4. **Write the answer:**
   Liam spends **$4.75** in total.
5. **Check:**
   Add again to confirm: $2.25 + $2.50 = $4.75 ✅

---

2. Converting Units Within the Same Measurement System
Example:
Convert 3 meters into centimeters.
Steps:
1. **Know the conversion:**
   1 meter = 100 centimeters
2. **Multiply:**
   $$3 \times 100 = 300$$
3. **Write the answer:**
   3 meters = **300 centimeters**
4. **Double-check:**
   Does 1 meter = 100 cm? Yes ✅
5. **Explain:**
   We multiplied because meters are larger than centimeters.

---

3. Understanding Area and Perimeter
Example:
A rectangle is 8 cm long and 5 cm wide. Find its area and perimeter.
Steps:
1. **Area formula:**
   $$\text{Area} = \text{length} \times \text{width}$$
   $$8 \times 5 = 40\ \text{cm}^2$$
2. **Perimeter formula:**
   $$\text{Perimeter} = 2(\text{length} + \text{width})$$
   $$2(8 + 5) = 2(13) = 26\ \text{cm}$$

3. **Write answers:**
   Area = **40 cm²**, Perimeter = **26 cm**
4. **Check:**
   Add sides: 8 + 5 + 8 + 5 = 26 ✅
5. **Explain:**
   Area covers inside space; perimeter measures boundary.

---

4. Understanding Types of Angles
Example:
Classify these angles: 45°, 90°, 120°.
Steps:

1. **Know definitions:**
   - Acute < 90°
   - Right = 90°
   - Obtuse > 90° but < 180°
2. **Check each angle:**
   - 45° → Acute
   - 90° → Right
   - 120° → Obtuse
3. **Write classifications:**
   45° = Acute, 90° = Right, 120° = Obtuse
4. **Double-check:**
   Compare each to 90° ✅
5. **Explain:**
   Acute = sharp, Right = square corner, Obtuse = wide.

---

5. Representing and Interpreting Data (Graphs & Pictographs)
Example:
Students voted for favorite fruit: Apples (8), Bananas (5), Grapes (7). Make a bar graph and find which fruit is most popular.
Steps:

1. **Draw axes:**
   - X-axis = Fruits
   - Y-axis = Number of votes
2. **Draw bars:**
   - Apples = 8
   - Bananas = 5
   - Grapes = 7
3. **Label graph:**
   Title: "Favorite Fruits"
4. **Interpret:**
   Which bar is tallest? Apples (8 votes)

Answer:

Apples are most popular.

Measurement & Data – Pretest

1. Lily has a ribbon that is 3 meters long. How many centimeters is the ribbon?

2. A watermelon weighs 4,500 grams. How many kilograms does it weigh?

3. A jug holds 2 liters of juice. Each glass holds 250 milliliters. How many full glasses can be filled?

4. A movie starts at 2:15 p.m. and ends at 4:00 p.m. How long is the movie?

5. Kalel has $20. He buys a toy for $8.75 and a snack for $3.50. How much money does he have left?

6. A rectangle has a length of 8 cm and a width of 5 cm. What is its perimeter?

7. What is the difference between an acute angle, a right angle, and a obtuse angle?

8. What type of angle is 120°? (acute, right, obtuse)

9. Line Plot Interpretation - Word Problem
The table below shows the number of books read by students in one month:

Books Read	Number of Students
1	2
2	3
3	4
4	5
5	3
6	1

Question:
How many students read more than 3 books?

10. Bar Graph Interpretation
Question:
The bar graph below shows the number of pets owned by students in a class:
- Dogs: 8 students
- Cats: 5 students
- Fish: 3 students
- Birds: 2 students

What animal is owned by students four times more than birds?

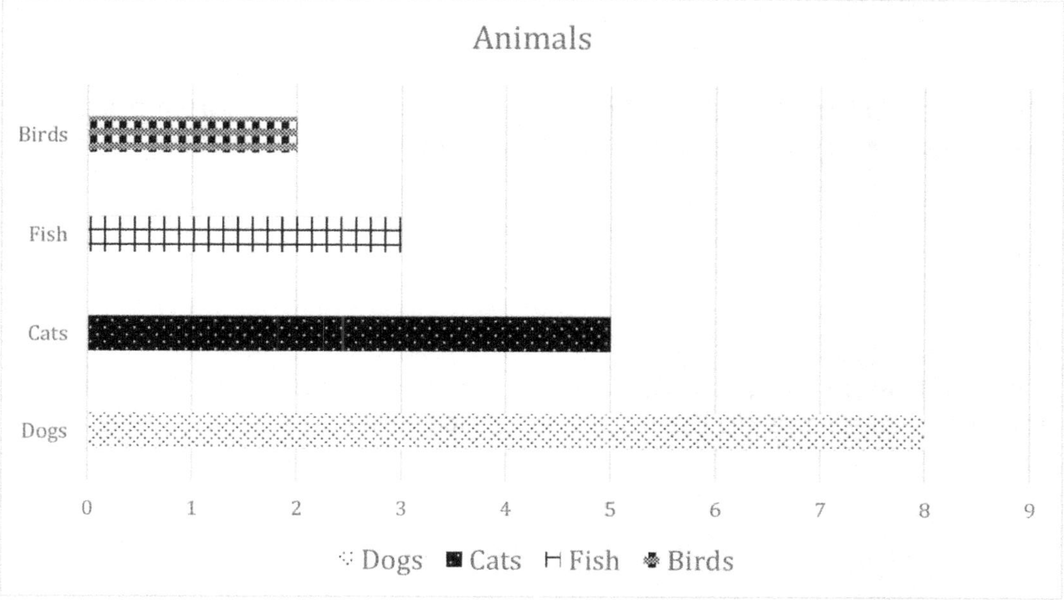

4th Grade Math – Solving Word Problems Using Length, Mass, Volume, Type, & Money

1. Evelynn ran 3 kilometers in the morning and 2,500 meters in the afternoon. How many meters did she run in total?

2. A piece of ribbon is 120 centimeters long. Gabby cuts it into 4 equal pieces. How long is each piece in centimeters?

3. A watermelon weighs 5 kilograms, and a pineapple weighs 2 kilograms. How many kilograms do both fruits weigh together?

4. A bag of sugar weighs 4,000 grams. If you use 1,250 grams for baking, How many grams of sugar are left in the bag?

5. A water bottle holds 600 milliliters. Liam fills 5 bottles completely. How many liters of water does he have in total?

6. A pitcher holds 2 liters of juice. Each glass holds 250 milliliters. How many glasses can be filled from one pitcher?

7. Sophia started reading at 2:45 p.m. and finished at 4:10 p.m. How long did she spend reading?

8. The movie started at 6:30 p.m. and lasted 1 hour 45 minutes. What time did the movie end?

9. Kalel bought 3 notebooks for $2.50 each and a pack of pencils for $1.75. How much did he spend in total?

10. Whitley had $20. She bought a book for $8.50 and a snack for $2.75. How much money does she have left?

11. A garden path is 18 meters long. Lily places a solar light every 3 meters along the path, including one at the start. How many lights does she need in total?

12. A train travels 425 kilometers on Monday and 375 kilometers on Tuesday. How many kilometers does it travel in both days combined?

13. A pumpkin weighs 3,200 grams. A watermelon weighs twice as much. What is the mass of the watermelon in kilograms?

14. Sam buys 5 bags of apples. Each bag weighs 2 kilograms. What is the total weight of the apples in grams?

15. A jug can hold 1.5 liters of milk. If you pour the milk into small bottles that each hold 250 milliliters, how many full bottles can you fill?

16. A fish tank holds 30 liters of water. After cleaning, only 12 liters remain. How many more liters of water are needed to fill the tank again?

17. The bus left at 7:25 a.m. and arrived at 9:10 a.m. How long was the bus ride?

18. A school play begins at 6:45 p.m. and ends at 8:05 p.m. How many minutes long is the play?

19. Armani bought 2 toy cars for $4.25 each and a board game for $12.50. What was his total cost?

20. Brett had $50. She spent $19.75 on a shirt and $8.60 on lunch. How much money does she have left?

21. A rope is 4 meters long. Jake cuts it into pieces that are each 80 centimeters long. How many pieces can he cut from the rope?

22. A cat weighs 6 kilograms. Her kitten weighs 2,500 grams. How much heavier is the cat than her kitten?

23. A pot holds 5 liters of soup. If each bowl holds 400 milliliters, How many full bowls of soup can be served from the pot?

24. School starts at 8:15 a.m. and ends at 3:00 p.m. How many hours and minutes are students in school each day?

25. Caroline has $25. She buys 3 notebooks for $4.75 each.
How much money will she have left after her purchase?

Solving Word Problems: Unit Conversion 1-25

1. Isabella measured a ribbon that was 3 meters long. How many centimeters long is the ribbon?

2. A racetrack is 2,500 meters long. How many kilometers is the racetrack?

3. A pencil is 18 centimeters long. How many millimeters is that?

4. A garden is 6 meters wide and 900 centimeters long. What is the length of the garden in meters?

5. A bag of rice weighs 4 kilograms. How many grams does it weigh?

6. A cat weighs 4,500 grams. How many kilograms is that?

7. A baker used 2 kilograms of flour to bake bread and 500 grams for cupcakes. How many grams of flour did the baker use in total?

8. A water jug holds 3 liters. How many milliliters of water does it hold?

9. A juice box contains 250 milliliters of juice. How many juice boxes are needed to make 2 liters of juice?

10. A movie lasts 2 hours and 30 minutes. How many minutes long is the movie?

11 A school hallway is 12 meters long. How many centimeters is that?

12. The distance from Mia's house to the park is 3.5 kilometers. How many meters is the distance?

13. A gymnastics mat is 800 centimeters long. How many meters is that?

14. A bike race is 5 kilometers long. If the riders have gone 2,000 meters, how many kilometers do they still need to ride?

15. A watermelon weighs 6,000 grams. How many kilograms is that?

16. A puppy gained 2 kilograms in one month. How many grams did it gain?

17. A sandwich weighs 250 grams. How many sandwiches would equal 1 kilogram?

18. A pitcher holds 4 liters of lemonade. How many milliliters does it hold?

19. Each small bottle holds 500 milliliters of water. How many bottles are needed to fill a 3-liter container?

20. A car trip lasted 2 hours and 45 minutes. How many minutes long was the trip?

21. A jump rope is 2.4 meters long. How many centimeters long is the jump rope?

22. A package weighs 3 kilograms and 250 grams. How many grams does the package weigh in total?

23. A magical fish tank holds 7,500 milliliters of water. How many liters of water is that?

24. Basketball practice lasts 1 hour and 20 minutes. How many minutes long is the practice?

25. Avery works for 3 hours and 45 minutes on Saturday. How many minutes does she work in total?

Understanding Area & Perimeter 1-20

1. Lily opens a book on her desk. The pages make an angle smaller than a right angle. What type of angle do the pages form?

2. A clock shows the time 3:00. The hands of the clock form an angle that looks like a perfect "L." What type of angle is shown?

3. Robert opens a door halfway so that it forms a wide space between the door and the wall — larger than a right angle but smaller than a straight line. What kind of angle did Robert make?

4. On the playground, a triangle is painted on the ground. One corner measures 120°. Is this angle acute, obtuse, or right?

5. Maria cuts a piece of paper into the shape of a triangle. One corner is 90°, one is 60°, and one is 30°. Name each type of angle in her triangle.

6. At the skate park, a ramp meets the ground at an angle less than 90°. What type of angle is formed between the ramp and the ground?

7. A window frame corner forms a perfect square corner. What type of angle is this, and how many degrees does it measure?

8. When the clock shows 10:00, the hands make an angle larger than a right angle. What type of angle is shown on the clock?

9. A triangle has one angle that measures 45°, one that measures 90°, and one that measures 45°. Which angles are acute, and which is right?

10. In art class, Noah draws two lines that cross and form a wide angle of 135°. What kind of angle did Noah draw?

11. Evy folds a piece of paper so the corners line up perfectly, forming an "L" shape. What type of angle did she make where the fold meets?

12. A painter leans a ladder against a wall. The angle between the ladder and the ground is smaller than 90°. What kind of angle is formed between the ladder and the ground?

13. In math class, Marco measures an angle that is 100° using a protractor. Is the angle acute, right, or obtuse?

14. The corner of a rectangular whiteboard forms a 90° angle. If the teacher draws a line inside the rectangle that slants up from the corner, making a 60° angle, what type of angle is that new line forming?

15. A triangle has one angle that measures 120°. What type of angle is that, and what does it tell you about the shape of the triangle?

16. When the clock shows 2:00, the hands make a small angle between them. Is the angle acute, right, or obtuse?

17. During art class, Ava draws a star. One of the points of the star measures 40°. What type of angle is that point of the star?

18. A carpenter cuts two pieces of wood so they meet to form a 120° angle for a shelf design. What kind of angle did the carpenter make?

19. The corner of a classroom floor and wall meet to form a perfect "L." What type of angle is formed there?

20. Ben draws two lines that cross. One of the angles formed measures 75°, and the other measures 105°. Identify which angle is acute and which is obtuse.

Identify the Type of Angle 1-20

1. 36° → ____________________

2. 78° → ____________________

3. 100° → ____________________

4. 180° → ____________________

5. 270° → ____________________

6. 45° → ____________________

7. 120° → ____________________

8. 90° → ____________________

9. 150° → ____________________

10. 15° → ____________________

11. 25° → ____________________

12. 60° → ____________________

13. 95° → ____________________

14. 135° → ____________________

15. 89° → ____________________

16. 200° → ____________________

17. 310° → ____________________

18. 170° → ____________________

19. 250° → ____________________

20. 80° → ____________________

Find the Perimeter 1-10

1. A rectangle (playground) has side lengths 26 m and 18 m.
   Find: The perimeter (total distance around).

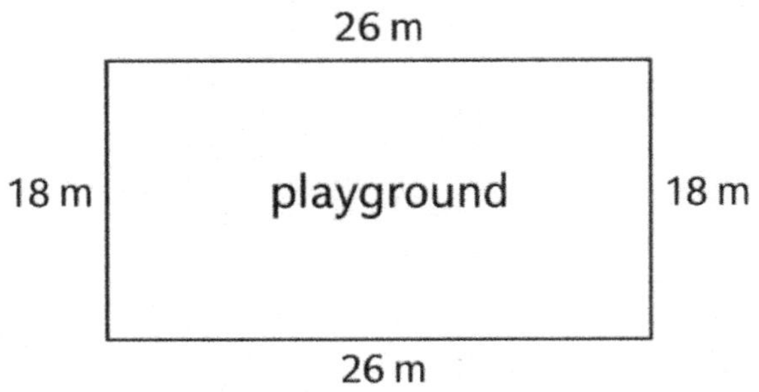

2. A trapezoid has side lengths 11m and 14m and angles of 45 and 120. What is the perimeter, and what are the types of angles?

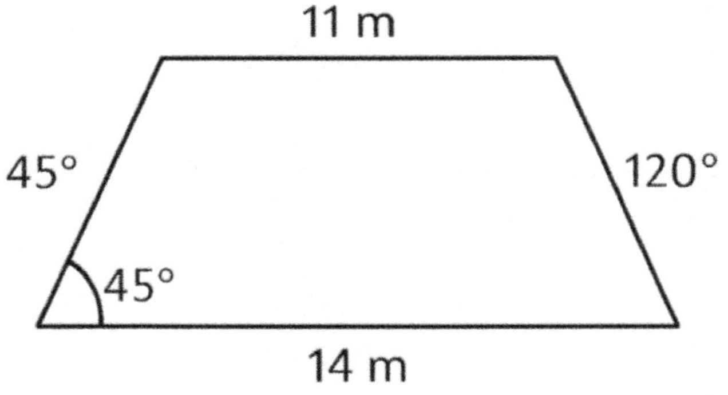

3. What is the perimeter of the pentagon?

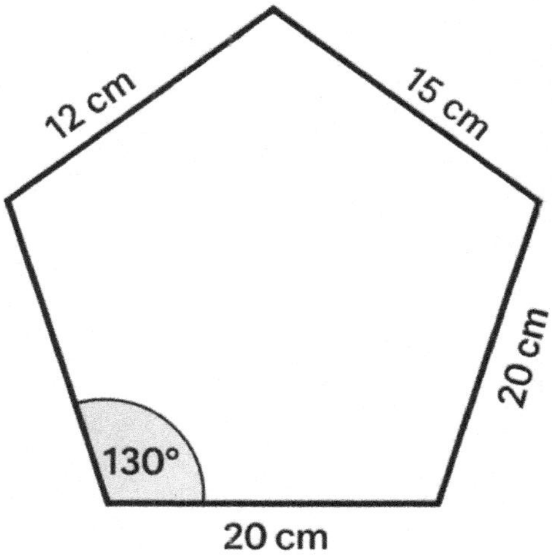

4. A rectangular swimming pool has a length of 32 meters and a width of 14 meters. Find the perimeter (total distance around the pool) AND find the area (space inside the pool).

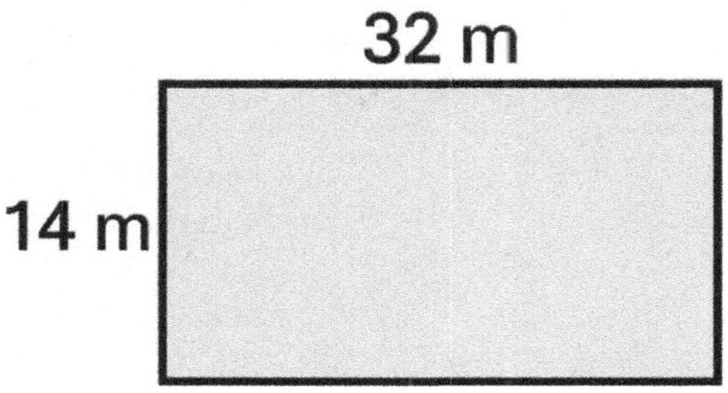

5. A hexagon (six-sided shape) has all sides equal to 9cm. Find the perimeter (total distance around the hexagon).

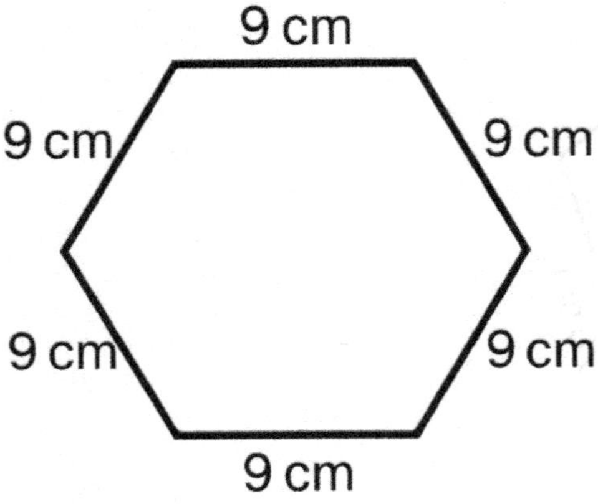

6. A rectangle has side lengths of 21m and 17 m. Find the perimeter and the area of the rectangle.

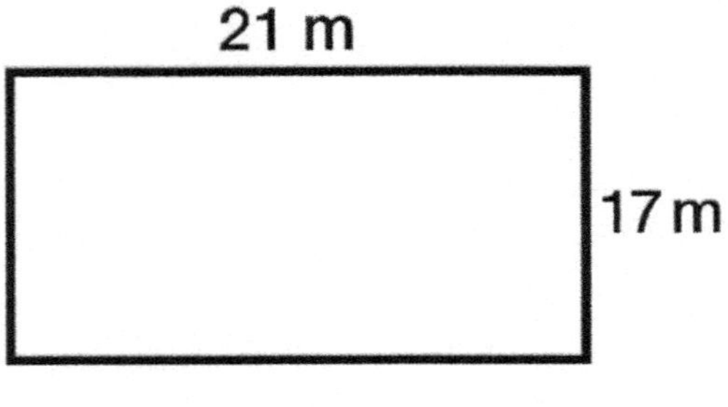

7. A rhombus (four-sided shape) with all sides measuring 12 inches. Find the perimeter of the rhombus.

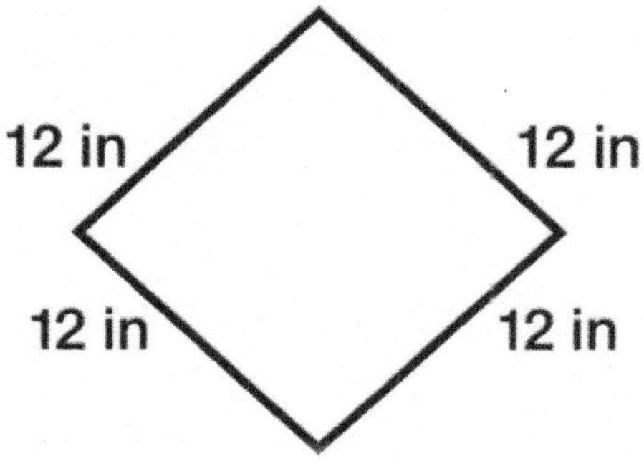

8. A triangle has side lengths of 25cm, 13cm, and 13cm. Find the perimeter of the triangle.

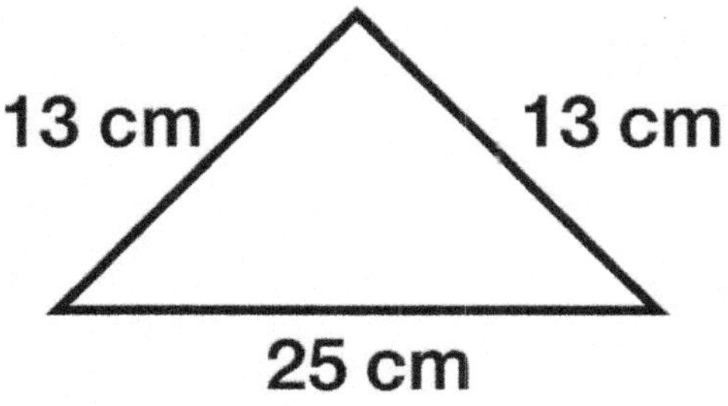

9. An octagon (eight-sided shape) has all sides measuring 6 meters. Can you find the perimeter of the octagon?

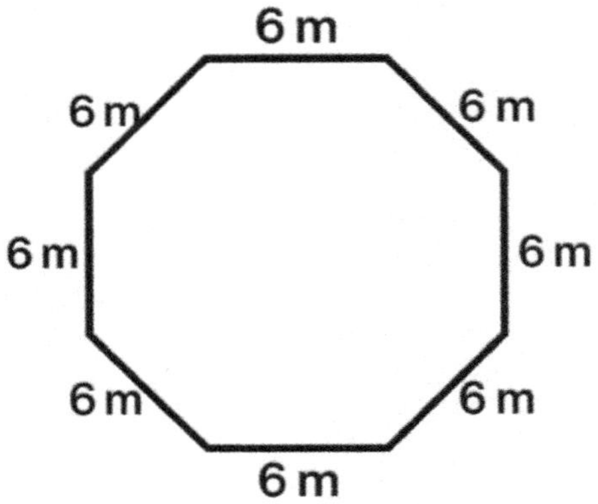

10. A parallelogram has two sides that are 28 feet and two sides that are 14 feet. Find the perimeter of the parallelogram.

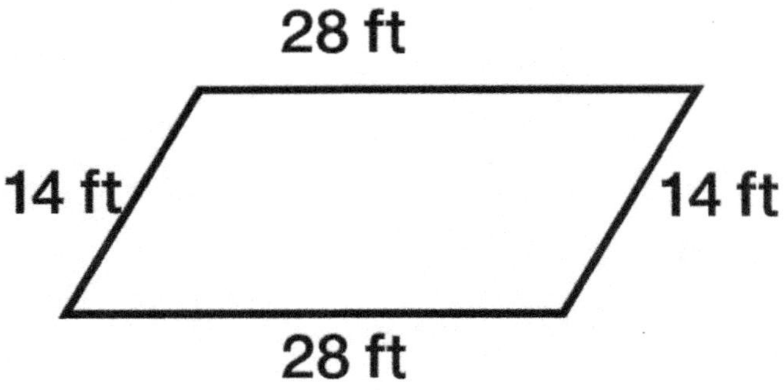

Convert the Measurements 1-30

1. Convert 4 meters (m) into centimeters (cm).

2. Convert 850 centimeters (cm) into meters (m).

3. Convert 5 kilometers (km) into meters (m).

4. Convert 2,000 millimeters (mm) into meters (m).

5. Convert 3 kilograms (kg) into grams (g).

6. Convert 7,500 grams (g) into kilograms (kg).

7. Convert 160 ounces (oz) into pounds (lb).

8. Convert 4 pounds (lb) into ounces (oz).

9. Convert 5 liters (L) into milliliters (mL).

10. Convert 2,500 milliliters (mL) into liters (L).

11. Convert 1.5 meters (m) into centimeters (cm).

12. Convert 6 feet (ft) into inches (in).

13. Convert 4 yards (yd) into feet (ft).

14. Convert 3 miles (mi) into feet (ft).

15. Convert 5 kilograms (kg) into grams (g).

16. Convert 4,500 grams (g) into kilograms (kg).

17. Convert 48 ounces (oz) into pounds (lb).

18. Convert 3 pounds (lb) into ounces (oz).

19. Convert 8 cups (c) into pints (pt).

20. Convert 3 quarts (qt) into cups (c).

21. Convert 10 millimeters (mm) into centimeters (cm)

22. Convert 5 meters (m) into centimeters (cm)

23. Convert 3 kilometers (km) into meters (m)

24. Convert 250 centimeters (cm) into meters (m)

25. Convert 2 liters (L) into milliliters (mL)

26. Convert 1,500 milliliters (mL) into liters (L)

27. Convert 190 pounds (lbs) into kilograms (kg)

28. Convert 60 minutes into hours

29. Convert 3 hours into minutes

30. Convert 7 days into hours

Representing and Interpreting Data

1: Favorite Pets

A survey shows these results for favorite pets in a class:

- Dogs: 10
- Cats: 7
- Fish: 5
- Birds: 3

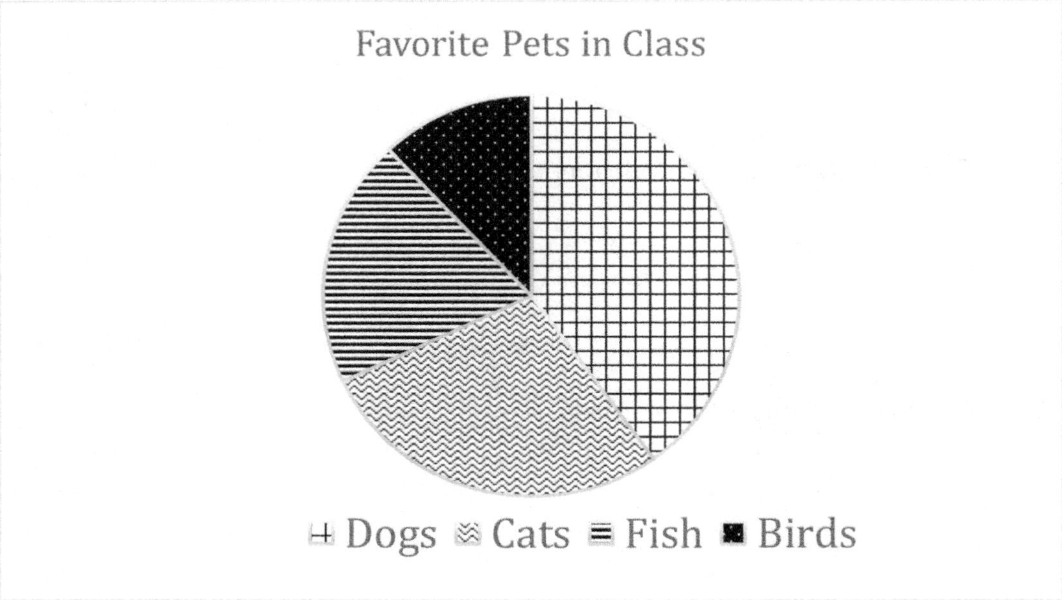

(a) Which pet is least popular?

(b) How many more students like dogs than birds?

2: Weekly Book Reading

Students read books in one week:

- Monday: 4 books
- Tuesday: 6 books
- Wednesday: 3 books
- Thursday: 5 books
- Friday: 7 books

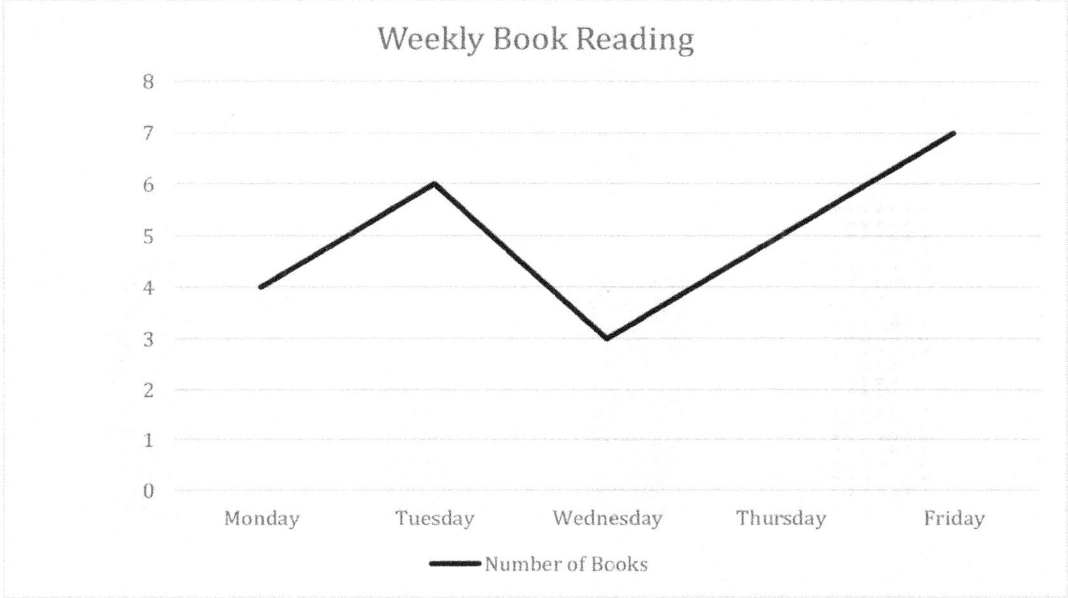

(a) On which day were the most books read?

(b) How

3: Fruit Sales

A fruit shop sold:

- Apples: 24
- Bananas: 18
- Grapes: 12
- Oranges: 20

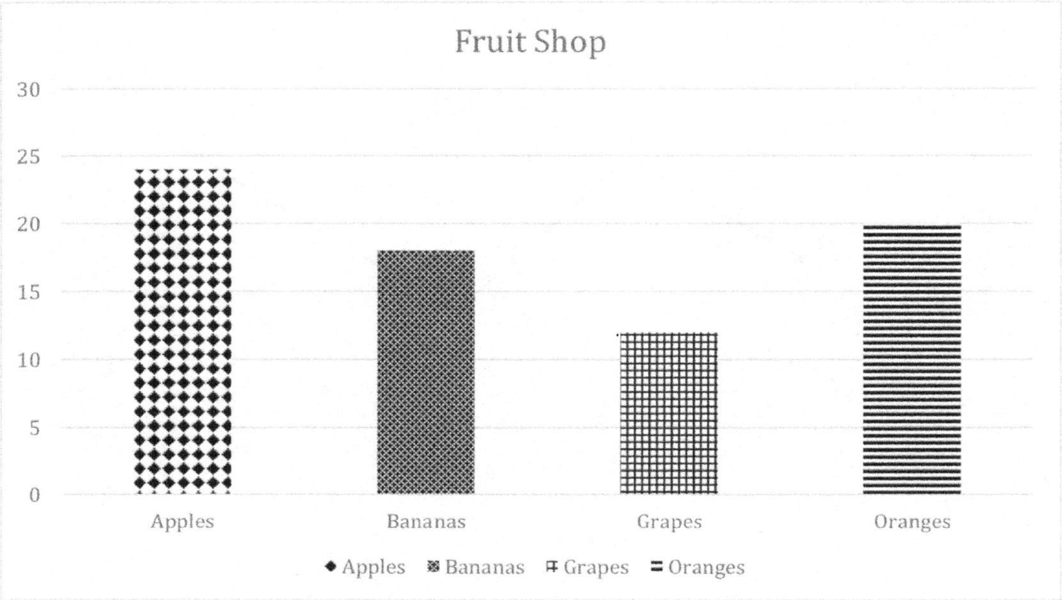

(a) Which fruit sold the most?

(b) How many more apples than grapes were sold?

4: Sports Club Members

Members in different sports:

- Soccer: 15
- Basketball: 10
- Tennis: 8
- Swimming: 12

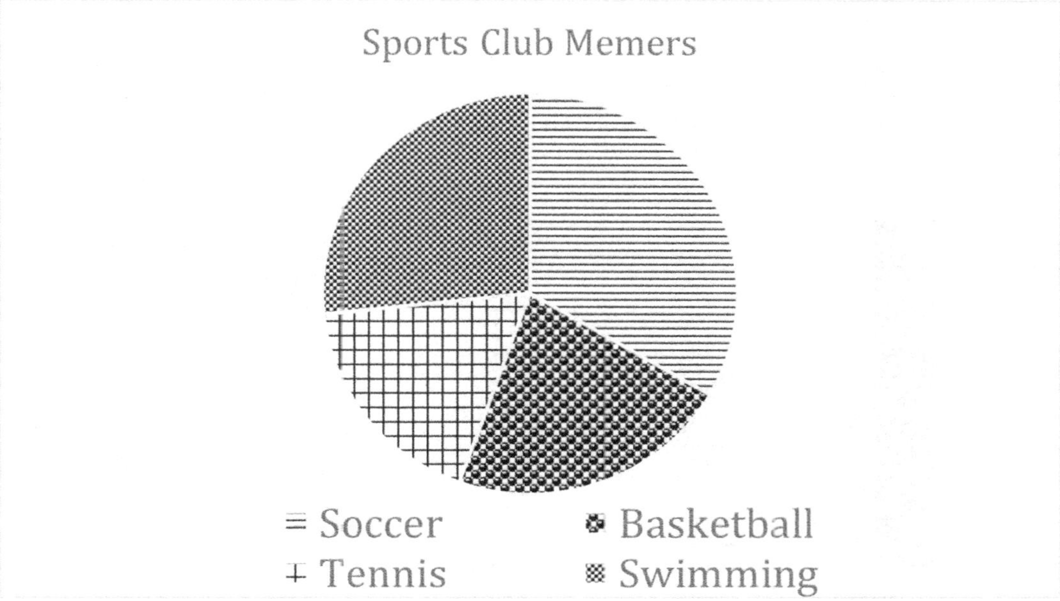

(a) Which sport has the second highest number of members?

(b) If 3 more members join Tennis, how many will Tennis have?

5: Favorite Ice Cream Flavors

Votes for flavors:

- Chocolate: 14
- Vanilla: 9
- Strawberry: 11
- Mint: 6

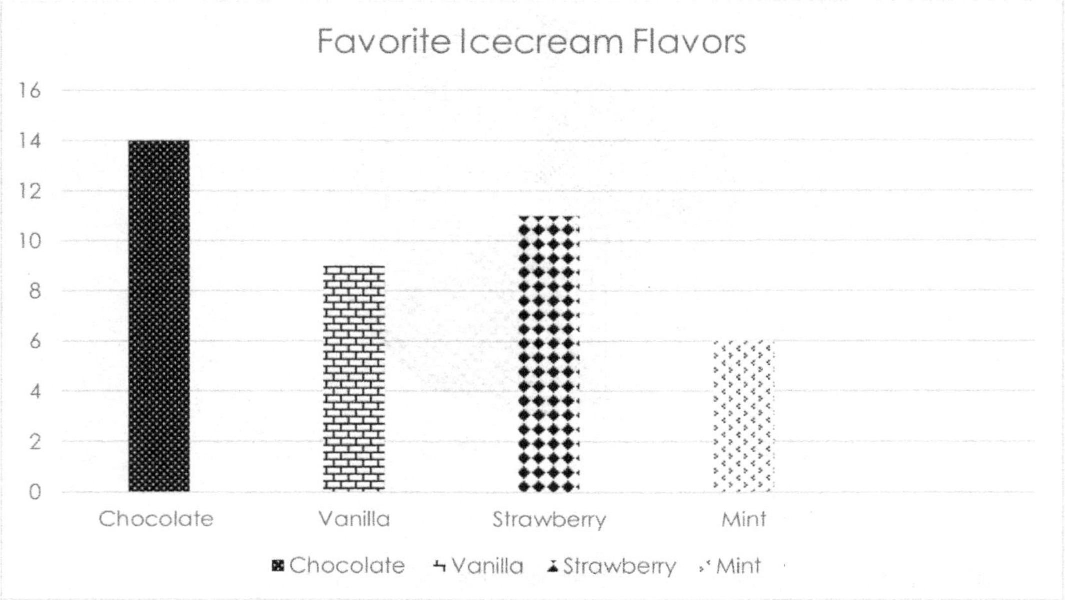

(a) Which flavor is most popular?

(b) How many total votes were cast?

Measurement and Data Post Test

Difficulty: Challenging

1. Composite Shape

A garden is shaped like an L.

- The long rectangle is 12 m by 4 m.
- The short rectangle is 4 m by 6 m.

(a) What is the total area of the garden?

(b) What is the perimeter of the garden?

2. Missing Side

A rectangular playground has an area of 96 m^2 and a perimeter of 40 m.
Find the length and width of the playground.

3. Angle Hunt in a Triangle

A triangle has three angles:

- Angle A = 65°
- Angle B = 90°
- Angle C = 25°

(a) Classify each angle as acute, right, or obtuse.

(b) Which type of triangle is this based on its angles?

4. Mixed Length & Time

Sofia is making a ribbon decoration. She has 3 pieces of ribbon:

- Piece A: 1.25 meters
- Piece B: 85 centimeters
- Piece C: 150 centimeters

She wants to join them into one long ribbon.

(a) What is the total length in meters?

(b) If she cuts the ribbon into 5 equal parts, how long is each part in centimeters?

5. Mixed Volume, Mass & Money

A juice shop sells orange juice in two sizes:

- Small bottle: 250 mL for $1.20
- Large bottle: 1.5 L for $5.40

A customer buys 3 small bottles and 2 large bottles.

(a) How many liters of juice does the customer buy?

(b) What is the total cost?

(c) If the customer pays with $20, how much change will they get?

6. Favorite Sports Survey

A class of 30 students voted for their favorite sport. The results are shown below:

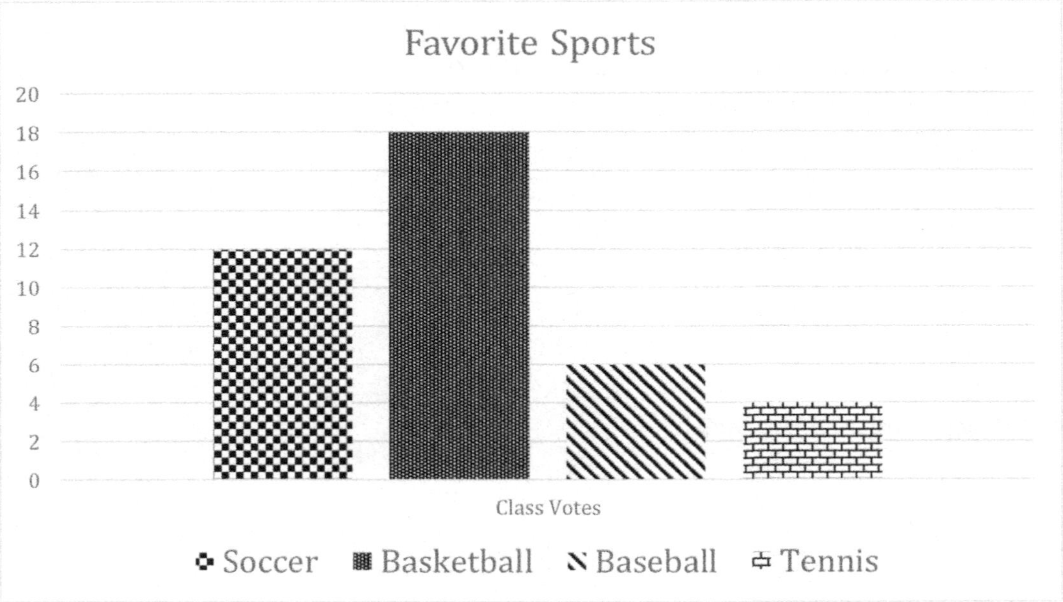

Questions:

(a) Which sport is the most popular?

(b) How many more students voted for Soccer than Tennis?

(c) If 5 more students voted for Basketball, how many votes would it have?

Answer Key – Step-by-Step Solutions

Pretest Solutions

1. Lily has a ribbon that is 3 meters long. How many centimeters is the ribbon?

Step 1: Understand the conversion factor: 1 meter = 100 centimeters
Step 2: Multiply the length in meters by the conversion factor: 3 meters × 100 = 300 centimeters

Answer: The ribbon is 300 centimeters long.

2. A watermelon weighs 4,500 grams. How many kilograms does it weigh?

Step 1: Understand the conversion factor: 1 kilogram = 1,000 grams
Step 2: Divide the weight in grams by the conversion factor: 4,500 ÷ 1,000 = 4.5 kilograms

Answer: The watermelon weighs 4.5 kilograms.

3. A jug holds 2 liters of juice. Each glass holds 250 milliliters. How many full glasses can be filled?

Step 1: Understand the conversion factor1 liter = 1,000 milliliters
Step 2: Convert liters to milliliters: 2 liters = 2 × 1,000 = 2,000 milliliters
Step 3: Divide the total juice by the volume of one glass: 2,000 ÷ 250 = 8

Answer: 8 full glasses can be filled.

4. A movie starts at 2:15 p.m. and ends at 4:00 p.m. How long is the movie?

Step 1: Break the time into parts: From 2:15 p.m. to 3:15 p.m. = 1 hour and from 3:15 p.m. to 4:00 p.m. = 45 minutes
Step 2: Add the time intervals: 1 hour + 45 minutes = 1 hour and 45 minutes

Answer: The movie is 1 hour and 45 minutes long.

5. Kalel has $20. He buys a toy for $8.75 and a snack for $3.50. How much money does he have left?

Step 1: Add the cost of the toy and the snack: $8.75 + $3.50 = $12.25
Step 2: Subtract the total cost from the amount Kalel had: $20.00 - $12.25 = $7.75

Answer: Kalel has $7.75 left.

7. Understanding Angles: Acute, Right, and Obtuse

Step 1: Understand what an angle is
An angle is formed when two lines (or rays) meet at a point called the vertex. The amount of turn between the two lines is measured in degrees (°).

Step 2: Define each type of angle

◆ **Acute Angle**
Definition: An angle that measures less than 90°.
Example: 45°, 30°, 75°
Visual: Looks like a sharp, narrow angle.

◆ **Right Angle**
Definition: An angle that measures exactly 90°.
Example: 90°
Visual: Forms a perfect "L" shape, like the corner of a square.

◆ **Obtuse Angle**
Definition: An angle that measures more than 90° but less than 180°.
Example: 120°, 135°, 100°
Visual: Appears wide or stretched out.

Step 3: Compare the angles

Type of Angle	Degree Range	Description
Acute	Less than 90°	Sharp and narrow
Right	Exactly 90°	Perfect corner (L-shape)
Obtuse	Between 90°–180°	Wide and open

Summary
✅ • **Acute = small angle (< 90°)**
✅ • **Right = perfect angle (90°)**
✅ • **Obtuse = large angle (> 90° but < 180°)**

8. What type of angle is 120°? (acute, right, obtuse)
Step 1: Recall angle types:
• Acute angle: Less than 90°
• Right angle: Exactly 90°
• Obtuse angle: Greater than 90° but less than 180°

Step 2: Compare 120° to these definitions:
• 120° is greater than 90°
• 120° is less than 180°

Step 3: Decide the type:
Since 120° is between 90° and 180°, it is an obtuse angle.
✅ **Final Answer: 120° is an obtuse angle.**

9. Bar Graph Interpretation - How many more students own dogs than cats?

Step 1. Identify the number of students who own dogs: 8

Step 2. Identify the number of students who own cats: 5

Step 3. Subtract the number of students who own cats from those who own dogs:

8 - 5 = 3

Answer: 3 more students own dogs than cats.

10. Bar Graph Interpretation - What animal is owned by students four times more than birds?

Step 1: Find the number of students who own birds.

- Birds = 2 students

Step 2: Multiply the number of bird owners by 4.

- 2 × 4 = 8 students

Step 3: Look at the data to see which animal is owned by 8 students.

- Dogs = 8 students

Step 4: The animal owned by 8 students is dogs.

Final Answer: Dogs are owned by students four times more than birds.

Solutions: Solving Word Problems Using Length, Mass, Volume, Time & Money: 1-25

1. Evelynn ran 3 kilometers in the morning and 2,500 meters in the afternoon. How many meters did she run in total?
Step 1: Convert kilometers to meters: 3 km × 1,000 = 3,000 meters
Step 2: Add both distances: 3,000 meters + 2,500 meters = 5,500 meters
✅ **Answer: Evelynn ran 5,500 meters in total.**

2. A piece of ribbon is 120 centimeters long. Gabby cuts it into 4 equal pieces. How long is each piece in centimeters?
Step 1: Divide the total length by the number of pieces: 120 ÷ 4 = 30 centimeters
✅ **Answer: Each piece is 30 centimeters long.**

3. A watermelon weighs 5 kilograms, and a pineapple weighs 2 kilograms. How many kilograms do both fruits weigh together?
Step 1: Add the weights: 5 kg + 2 kg = 7 kilograms
✅ **Answer: Both fruits weigh 7 kilograms together.**

4. A bag of sugar weighs 4,000 grams. If you use 1,250 grams for baking,
How many grams of sugar are left in the bag?
Step 1: Subtract the amount used from the total: 4,000 − 1,250 = 2,750 grams
✅ **Answer: 2,750 grams of sugar are left in the bag.**

5. A water bottle holds 600 milliliters. Liam fills 5 bottles completely.
How many liters of water does he have in total?
Step 1: Find the total milliliters: 600 × 5 = 3,000 milliliters
Step 2: Convert milliliters to liters: 3,000 ÷ 1,000 = 3 liters
✅ **Answer: Liam has 3 liters of water in total.**

6. A pitcher holds 2 liters of juice. Each glass holds 250 milliliters. How many glasses can be filled from one pitcher?
Step 1: Convert liters to milliliters: 1 liter = 1,000 milliliters
Step 2: 2 liters = 2 × 1,000 = 2,000 milliliters
Step 3: Divide total juice by the capacity of one glass: 2,000 ÷ 250 = 8 glasses
✅ **Answer: 8 glasses can be filled from one pitcher.**

7. Sophia started reading at 2:45 p.m. and finished at 4:10 p.m. How long did she spend reading?
Step 1: Calculate the time difference: From 2:45 to 3:45 = 1 hour
Step 2: From 3:45 to 4:10 = 25 minutes
Step 3: Add both durations: 1 hour + 25 minutes = 1 hour 25 minutes
✅ **Answer: Sophia spent 1 hour and 25 minutes reading.**

8. The movie started at 6:30 p.m. and lasted 1 hour 45 minutes. What time did the movie end?
Step 1: Add the duration to the start time: 6:30 + 1 hour = 7:30
Step 2: 7:30 + 45 minutes = 8:15
✅ **Answer: The movie ended at 8:15 p.m.**

9. Kalel bought 3 notebooks for $2.50 each and a pack of pencils for $1.75. How much did he spend in total?
Step 1: Calculate the cost of notebooks: 3 × $2.50 = $7.50
Step 2: Add the cost of pencils: $7.50 + $1.75 = $9.25
✅ **Answer: Kalel spent $9.25 in total.**

10. Whitley had $20. She bought a book for $8.50 and a snack for $2.75. How much money does she have left?
Step 1: Add the expenses: $8.50 + $2.75 = $11.25
Step 2: Subtract from the total amount: $20.00 – $11.25 = $8.75
✅ **Answer: Whitley has $8.75 left.**

11. A garden path is 18 meters long. Lily places a solar light every 3 meters along the path, including one at the start. How many lights does she need in total?
Step 1: The first light is placed at the start (0 meters).
Step 2: Lights are placed every 3 meters along the 18-meter path.
Step 3: Divide the total length by the spacing: 18 ÷ 3 = 6.
Step 4: Including the light at the start, total lights = 6 + 1 = 7.
✅ **Answer: Lily needs 7 solar lights in total.**

12. A train travels 425 kilometers on Monday and 375 kilometers on Tuesday. How many kilometers does it travel in both days combined?
Step 1: Add the distances traveled on both days.
Step 2: 425 km + 375 km = 800 km.
✅ **Answer: The train travels 800 kilometers in total.**

13. A pumpkin weighs 3,200 grams. A watermelon weighs twice as much. What is the mass of the watermelon in kilograms?
Step 1: Multiply the pumpkin's weight by 2 to find the watermelon's weight.
Step 2: 3,200 × 2 = 6,400 grams.
Step 3: Convert grams to kilograms: 6,400 ÷ 1,000 = 6.4 kg.
✅ **Answer: The watermelon weighs 6.4 kilograms.**

14. Sam buys 5 bags of apples. Each bag weighs 2 kilograms. What is the total weight of the apples in grams?
Step 1: Multiply the number of bags by the weight of each bag.
Step 2: 5 × 2 = 10 kilograms.
Step 3: Convert kilograms to grams: 10 × 1,000 = 10,000 grams.
✅ **Answer: The total weight of the apples is 10,000 grams.**

15. A jug can hold 1.5 liters of milk. If you pour the milk into small bottles that each hold 250 milliliters, how many full bottles can you fill?
Step 1: Convert liters to milliliters: 1.5 × 1,000 = 1,500 milliliters.
Step 2: Divide the total milk by the capacity of each bottle: 1,500 ÷ 250 = 6.
Answer: You can fill 6 full bottles.

16. A fish tank holds 30 liters of water. After cleaning, only 12 liters remain. How many more liters of water are needed to fill the tank again?
Step 1: Identify the total capacity of the fish tank: The tank can hold 30 liters of water.
Step 2: Determine how much water is currently in the tank: After cleaning, only 12 liters remain.
Step 3: Subtract the current amount of water from the total capacity to find out how much more is needed: 30 liters (total capacity) - 12 liters (current amount) = 18 liters
Answer: 18 liters

17. The bus left at 7:25 a.m. and arrived at 9:10 a.m. How long was the bus ride? Start time: 7:25 a.m.
Step 1: First, calculate the minutes from 7:25 a.m. to 8:00 a.m.:
8:00 a.m. - 7:25 a.m. = 35 minutes
Step 2: Then calculate the time from 8:00 a.m. to 9:10 a.m.:
1 hour and 10 minutes
Step 3: Add both time intervals together:
35 minutes + 1 hour 10 minutes = 1 hour and 45 minutes
Answer: 1 hour 45 minutes

18. A school play begins at 6:45 p.m. and ends at 8:05 p.m. How many minutes long is the play?
Step 1: Break the time into two parts
From 6:45 p.m. to 7:45 p.m. = 1 hour = 60 minutes
From 7:45 p.m. to 8:05 p.m. = 20 minutes
Step 2: Add the two time intervals: 60 minutes + 20 minutes = 80 minutes
Answer: 80 minutes

19. Armani bought 2 toy cars for $4.25 each and a board game for $12.50. What was his total cost?
Step 1: Find the cost of toy cars = 2 × $4.25 = $8.50
Step 2: Find the cost of board game = $12.50
Step 3: Calculate the total cost = $8.50 + $12.50 = $21.00
Answer: $21.00

20. Brett had $50. She spent $19.75 on a shirt and $8.60 on lunch. How much money does she have left?
Step 1: Find the total spent = $19.75 + $8.60 = $28.35
Step 2: Calculate the Remaining = $50 – $28.35 = $21.65
Answer: $21.65

21. A rope is 4 meters long. Jake cuts it into pieces that are each 80 centimeters long. How many pieces can he cut from the rope?

Step 1: Convert 4 meters to centimeters: 4 × 100 = 400 cm
Step 2: Number of pieces = 400 ÷ 80 = 5
Answer: 5 pieces

22. A cat weighs 6 kilograms. Her kitten weighs 2,500 grams. How much heavier is the cat than her kitten?

Step 1: Convert 6 kg to grams: 6 × 1,000 = 6,000 grams
Step 2: Find the Difference = 6,000 − 2,500 = 3,500 grams
Answer: 3,500 grams

23. A pot holds 5 liters of soup. If each bowl holds 400 milliliters, how many full bowls of soup can be served from the pot?

Step 1: Convert 5 liters to milliliters: 5 × 1,000 = 5,000 ml
Step 2: Number of bowls = 5,000 ÷ 400 = 12.5 → 12 full bowls
Answer: 12 full bowls

24. School starts at 8:15 a.m. and ends at 3:00 p.m. How many hours and minutes are students in school each day?

Step 1: Start time: 8: Step 1: Break the time into two parts
From 8:15 a.m. to 12:00 p.m. = 3 hours and 45 minutes
From 12:00 p.m. to 3:00 p.m. = 3 hours
Step 2: Add both time intervals: 3 hours 45 minutes + 3 hours = 6 hours and 45
Answer: Students are in school for 6 hours and 45 minutes each day.

25. Caroline has $25. She buys 3 notebooks for $4.75 each. How much money will she have left after her purchase?

Step 1: Cost of notebooks = 3 × $4.75 = $14.25
Step 2: Remaining = $25 − $14.25 = $10.75
Answer: $10.75

Solutions: Solving Word Problems: Unit Conversion: 1-25

1. Isabella measured a ribbon that was 3 meters long. How many centimeters long is the ribbon?
Step 1: 1 meter = 100 centimeters– *Facts!*
Step 2: 3 meters × 100 = 300 centimeters
✅ **Answer: The ribbon is 300 centimeters long.**

2. A racetrack is 2,500 meters long. How many kilometers is the racetrack?
Step 1: 1 kilometer = 1,000 meters– *Facts!*
Step 2: 2,500 ÷ 1,000 = 2.5 kilometers
✅ **Answer: The racetrack is 2.5 kilometers long.**

3. A pencil is 18 centimeters long. How many millimeters is that?
Step 1: 1 centimeter = 10 millimeters– *Facts!*
Step 2: 18 × 10 = 180 millimeters
✅ **Answer: The pencil is 180 millimeters long.**

4. A garden is 6 meters wide and 900 centimeters long. What is the length of the garden in meters?
Step 1: 1 meter = 100 centimeters– *Facts!*
Step 2: 900 ÷ 100 = 9 meters
✅ **Answer: The garden is 9 meters long.**

5. A bag of rice weighs 4 kilograms. How many grams does it weigh?
Step 1: 1 kilogram = 1,000 grams– *Facts!*
Step 2: 4 × 1,000 = 4,000 grams
✅ **Answer: The bag of rice weighs 4,000 grams.**

6. A cat weighs 4,500 grams. How many kilograms is that?
Step 1: 1 kilogram = 1,000 grams– *Facts!*
Step 2: 4,500 ÷ 1,000 = 4.5 kilograms
✅ **Answer: The cat weighs 4.5 kilograms.**

7. A baker used 2 kilograms of flour to bake bread and 500 grams for cupcakes. How many grams of flour did the baker use in total?
Step 1: 2 kilograms = 2 × 1,000 = 2,000 grams
Step 2: 2,000 + 500 = 2,500 grams
✅ **Answer: The baker used 2,500 grams of flour in total.**

8. A water jug holds 3 liters. How many milliliters of water does it hold?
Step 1: 1 liter = 1,000 milliliters– *Facts!*
Step 2: 3 × 1,000 = 3,000 milliliters
Answer: The jug holds 3,000 milliliters of water.

9. A juice box contains 250 milliliters of juice. How many juice boxes are needed to make 2 liters of juice?
Step 1: 2 liters = 2 × 1,000 = 2,000 milliliters
Step 2: 2,000 ÷ 250 = 8 juice boxes
Answer: 8 juice boxes are needed to make 2 liters of juice.

10. A movie lasts 2 hours and 30 minutes. How many minutes long is the movie?
Step 1: 1 hour = 60 minutes– *Facts!*
Step 2: 2 × 60 = 120 minutes
120 + 30 = 150 minutes
Answer: The movie is 150 minutes long.

11. A school hallway is 12 meters long. How many centimeters is that?
Step 1: 1 meter = 100 centimeters– *Facts!*
Step 2: 12 meters × 100 = 1,200 centimeters
Answer: 1,200 centimeters

12. The distance from Mia's house to the park is 3.5 kilometers. How many meters is the distance?
Step 1: 1 kilometer = 1,000 meters– *Facts!*
Step 2: 3.5 kilometers × 1,000 = 3,500 meters
Answer: 3,500 meters

13. A gymnastics mat is 800 centimeters long. How many meters is that?
Step 1: 1 meter = 100 centimeters– *Facts!*
Step 2: 800 ÷ 100 = 8 meters
Answer: 8 meters

14. A bike race is 5 kilometers long. If the riders have gone 2,000 meters, how many kilometers do they still need to ride?
Step 1: Convert 2,000 meters to kilometers: 2,000 ÷ 1,000 = 2 kilometers
Step 2: 5 kilometers − 2 kilometers = 3 kilometers
Answer: 3 kilometers

15. A watermelon weighs 6,000 grams. How many kilograms is that?
Step 1: 1 kilogram = 1,000 grams– *Facts!*
Step 2: 6,000 ÷ 1,000 = 6 kilograms
Answer: 6 kilograms

16. A puppy gained 2 kilograms in one month. How many grams did it gain?
Step 1: 1 kilogram = 1,000 grams– *Facts!*
Step 2: 2 × 1,000 = 2,000 grams
Answer: 2,000 grams

17. A sandwich weighs 250 grams. How many sandwiches would equal 1 kilogram?
Step 1: 1 kilogram = 1,000 grams– *Facts!*
Step 2: 1,000 ÷ 250 = 4 sandwiches
Answer: 4 sandwiches

18. A pitcher holds 4 liters of lemonade. How many milliliters does it hold?
Step 1: 1 liter = 1,000 milliliters– *Facts!*
Step 2: 4 × 1,000 = 4,000 milliliters
Answer: 4,000 milliliters

19. Each small bottle holds 500 milliliters of water. How many bottles are needed to fill a 3-liter container?
Step 1: Convert 3 liters to milliliters: 3 × 1,000 = 3,000 milliliters
Step 2: 3,000 ÷ 500 = 6 bottles
Answer: 6 bottles

20. A car trip lasted 2 hours and 45 minutes. How many minutes was the trip?
Step 1: 1 hour = 60 minutes – *Facts!*
Step 2: 2 × 60 = 120 minutes
Step 3: 120 + 45 = 165 minutes
Answer: 165 minutes

20. A car trip lasted 2 hours and 45 minutes. How many minutes was the trip?
Step 1: Convert hours to minutes: 2 hours × 60 = 120 minutes
Step 2: Add the remaining 45 minutes: 120 + 45 = 165 minutes
Answer: The trip lasted 165 minutes.

21. A jump rope is 2.4 meters long. How many centimeters long is the jump rope?
Step 1: Convert meters to centimeters: 1 meter = 100 centimeters – *Facts!*
Step 2: 2.4 meters × 100 = 240 centimeters
Answer: The jump rope is 240 centimeters long.

22. A package weighs 3 kilograms and 250 grams. How many grams does the package weigh in total?
Step 1: Convert kilograms to grams: 3 kg × 1,000 = 3,000 grams
Step 2: Add the remaining 250 grams: 3,000 + 250 = 3,250 grams
Answer: The package weighs 3,250 grams in total.

23. A magical fish tank holds 7,500 milliliters of water. How many liters of water is that?

Step 1: Convert milliliters to liters: 1 liter = 1,000 milliliters – *Facts!*
Step 2: 7,500 ÷ 1,000 = 7.5 liters
☑ **Answer: The fish tank holds 7.5 liters of water.**

24. Basketball practice lasts 1 hour and 20 minutes. How many minutes is the practice?

Step 1: Convert hours to minutes: 1 hour × 60 = 60 minutes – *Facts!*
Step 2: Add the remaining 20 minutes: 60 + 20 = 80 minutes
☑ **Answer: Basketball practice lasts 80 minutes.**

25. Avery works for 3 hours and 45 minutes on Saturday. How many minutes does she work in total?

Step 1: Convert hours to minutes: 3 hours × 60 = 180 minutes
Step 2: Add the remaining 45 minutes: 180 + 45 = 225 minutes
☑ **Answer: Avery works for 225 minutes in total.**

Solutions: Understanding Area & Perimeter: 1-20

1. Lily opens a book on her desk. The pages make an angle smaller than a right angle. What type of angle do the pages form?
-A right angle measures exactly 90 degrees.
-An angle smaller than 90 degrees is called an acute angle.
-Since the pages form an angle smaller than a right angle, it is an acute angle.
✅ **Answer: Acute Angle**

2. A clock shows the time 3:00. The hands of the clock form an angle that looks like a perfect "L." What type of angle is shown?
-A perfect 'L' shape represents a 90-degree angle.
-An angle that measures exactly 90 degrees is called a right angle.
-Therefore, the angle formed by the clock hands is a right angle.
✅ **Answer: Right Angle**

3. Robert opens a door halfway so that it forms a wide space between the door and the wall — larger than a right angle but smaller than a straight line. What kind of angle did Robert make?
-A right angle is 90 degrees, and a straight angle is 180 degrees.
-An angle between 90 and 180 degrees is called an obtuse angle.
-Since the door forms an angle larger than 90 but less than 180 degrees, it is an obtuse angle.
✅ **Answer: Obtuse Angle**

4. On the playground, a triangle is painted on the ground. One corner measures 120°. Is this angle acute, obtuse, or right?
-An acute angle is less than 90 degrees.
-A right angle is exactly 90 degrees.
-An obtuse angle is greater than 90 degrees but less than 180 degrees.
-Since 120° is greater than 90°, it is an obtuse angle.
✅ **Answer: Obtuse Angle**

5. Maria cuts a piece of paper into the shape of a triangle. One corner is 90°, one is 60°, and one is 30°. Name each type of angle in her triangle.
-90° is a right angle.
-60° and 30° are both less than 90°, so they are acute angles.
-Therefore, the triangle has one right angle and two acute angles.
✅ **Answer: One Right Angle, Two Acute Angles**

6. At the skate park, a ramp meets the ground at an angle less than 90°. What type of angle is formed between the ramp and the ground?
-An angle less than 90° is called an acute angle.
-Since the ramp meets the ground at an angle less than 90°, it forms an acute angle.
Answer: Acute angle

7. A window frame corner forms a perfect square corner. What type of angle is this, and how many degrees does it measure?
-A perfect square corner measures exactly 90°.
-An angle of 90° is called a right angle.
Answer: Right angle (90°)

8. When the clock shows 10:00, the hands make an angle larger than a right angle. What type of angle is shown on the clock?
-A right angle is 90°.
-An angle larger than 90° but less than 180° is called an obtuse angle.
-At 10:00, the clock hands form an obtuse angle.
Answer: Obtuse angle

9. A triangle has one angle that measures 45°, one that measures 90°, and one that measures 45°. Which angles are acute, and which is right?
-Angles less than 90° are acute.
-An angle of 90° is a right angle.
-So, the 45° angles are acute, and the 90° angle is right.
Answer: 45° angles are acute; 90° angle is right

10. In art class, Noah draws two lines that cross and form a wide angle of 135°. What kind of angle did Noah draw?
-An angle greater than 90° but less than 180° is called an obtuse angle.
-135° is greater than 90°, so it is an obtuse angle.
Answer: Obtuse angle

11. Evy folds a piece of paper so the corners line up perfectly, forming an "L" shape. What type of angle did she make where the fold meets?
-An "L" shape forms a 90° angle.
-An angle of 90° is called a right angle.
Answer: Right angle

12. A painter leans a ladder against a wall. The angle between the ladder and the ground is smaller than 90°. What kind of angle is formed between the ladder and the ground?
-An angle smaller than 90° is called an acute angle.
-So, the angle between the ladder and the ground is acute.
Answer: Acute angle

13. In math class, Marco measures an angle that is 100° using a protractor. Is the angle acute, right, or obtuse?

-An angle greater than 90° but less than 180° is called an obtuse angle.
-100° is greater than 90°, so it is an obtuse angle.

Answer: Obtuse angle

14. The corner of a rectangular whiteboard forms a 90° angle. If the teacher draws a line inside the rectangle that slants up from the corner, making a 60° angle, what type of angle is that new line forming?

-An angle of 60° is less than 90°, so it is an acute angle.
-The new line forms an acute angle with the base of the rectangle.

Answer: Acute angle

15. A triangle has one angle that measures 120°. What type of angle is that, and what does it tell you about the shape of the triangle?

-An angle greater than 90° is called an obtuse angle.
-120° is greater than 90°, so it is an obtuse angle.
This tells us the triangle is an obtuse triangle.

Answer: Obtuse angle; the triangle is an obtuse triangle

16. When the clock shows 2:00, the hands make a small angle between them. Is the angle acute, right, or obtuse?

At 2:00, the hour hand points at 2 and the minute hand points at 12.
-The angle between them is less than 90°.
-An angle less than 90° is called an acute angle.

Answer: Acute angle

17. During art class, Ava draws a star. One of the points of the star measures 40°. What type of angle is that point of the star?

-The angle at the point of the star is 40°.
-Since 40° is less than 90°, it is an acute angle.

Answer: Acute angle

18. A carpenter cuts two pieces of wood so they meet to form a 120° angle for a shelf design. What kind of angle did the carpenter make?

-The angle formed is 120°.
-An angle greater than 90° but less than 180° is called an obtuse angle.

Answer: Obtuse angle

19. The corner of a classroom floor and wall meet to form a perfect "L." What type of angle is formed there?

-A perfect 'L' shape represents a 90° angle.
-An angle of 90° is called a right angle.

Answer: Right angle

20. Ben draws two lines that cross. One of the angles formed measures 75°, and the other measures 105°. Identify which angle is acute and which is obtuse.

-The angle measuring 75° is less than 90°, so it is an acute angle.

-The angle measuring 105° is greater than 90°, so it is an obtuse angle.

✅ **Answer: 75° is acute, 105° is obtuse**

Solutions: Identifying the Type of Angle: 1-30

1. 36° → Acute
2. 78° → Acute
3. 100° → Obtuse
4. 180° → Straight
5. 270° → Reflex
6. 45° → Acute
7. 120° → Obtuse
8. 90° → Right
9. 150° → Obtuse
10. 15° → Acute
11. 25° → Acute
12. 60° → Acute
13. 95° → Obtuse
14. 135° → Obtuse
15. 89° → Acute
16. 200° → Reflex
17. 310° → Reflex
18. 170° → Obtuse
19. 250° → Reflex
20. 80° → Acute

Solutions: Find the Perimeter Problems: 1-10

1. Rectangle (Playground): 26 m and 18 m
What we know:
- Length = 26 m
- Width = 18 m
Formula for Perimeter: P = 2 × (Length + Width)
Step 1: Add the length and width: 26 + 18 = 44
Step 2: Multiply by 2: 44 × 2 = 88 m
✅ **Answer: Perimeter = 88 meters**

2. Trapezoid: sides 11 m and 14 m, angles 45° and 120°
What we know:
- Two sides: 11 m and 14 m
- Angles: 45° and 120°
Step 1: Assume trapezoid is isosceles (two equal sides). Other two sides are also 11 m and 14 m.
Step 2: Add all sides: 11 + 14 + 11 + 14 = 50 m
Step 3: Identify angle types: 45° = acute angle, 120° = obtuse angle
✅ **Answer: Perimeter = 50 meters; Angle types: acute and obtuse**

3. Pentagon (assume all sides equal to 10 m)
What we know:
- 5 sides, each 10 m
Formula: P = side length × number of sides
Step 1: Multiply: 10 × 5 = 50 m
✅ **Answer: Perimeter = 50 meters**

4. Swimming Pool: 32 m and 14 m
What we know:
- Length = 32 m
- Width = 14 m
Perimeter Formula: P = 2 × (Length + Width)
Step 1: Add: 32 + 14 = 46
Step 2: Multiply by 2: 46 × 2 = 92 m
Area Formula: A = Length × Width
Step 3: Multiply: 32 × 14 = 448 m^2
✅ **Answer: Perimeter = 92 meters, Area = 448 m^2**

5. Hexagon: all sides 9 cm
What we know:
- 6 sides, each 9 cm
Formula: P = side length × number of sides
Step 1: Multiply: 9 × 6 = 54 cm
✅ **Answer: Perimeter = 54 cm**

6. Rectangle: 21 m and 17 m
What we know:
- Length = 21 m
- Width = 17 m

Perimeter Formula: P = 2 × (Length + Width)
Step 1: Add the sides: 21 + 17 = 38
Step 2: Multiply by 2: 38 × 2 = 76 m
Area Formula: A = Length × Width
Step 3: Multiply: 21 × 17 = 357 m^2
✅ **Answer: Perimeter = 76 m, Area = 357 m^2**

7. Rhombus: all sides 12 in
What we know:
- 4 equal sides = 12 in each

Perimeter Formula: P = side length × 4
Step 1: Multiply: 12 × 4 = 48 in
✅ **Answer: Perimeter = 48 inches**

8. Triangle: 25 cm, 13 cm, 13 cm
What we know:
- Three sides: 25 cm, 13 cm, 13 cm

Perimeter Formula: P = sum of all sides
Step 1: Add: 25 + 13 + 13 = 51 cm
✅ **Answer: Perimeter = 51 cm**

9. Octagon: all sides 6 m
What we know:
- 8 sides, each 6 m

Perimeter Formula: P = side length × number of sides
Step 1: Multiply: 6 × 8 = 48 m
✅ **Answer: Perimeter = 48 m**

10. Parallelogram: sides 28 ft and 14 ft
What we know:
- Two sides = 28 ft
- Two sides = 14 ft

Perimeter Formula: P = 2 × (sum of adjacent sides)
Step 1: Add adjacent sides: 28 + 14 = 42
Step 2: Multiply by 2: 42 × 2 = 84 ft
✅ **Answer: Perimeter = 84 ft**

Solutions: Convert the Measurements: 1-30

1. Convert 4 meters (m) into centimeters (cm)
What we know:
- 1 meter = 100 centimeters
Step 1: Multiply by 100: 4 × 100 = 400 cm
Answer: 4 m = 400 cm

2. Convert 850 centimeters (cm) into meters (m)
What we know:
- 1 meter = 100 centimeters
Step 1: Divide by 100: 850 ÷ 100 = 8.5 m
Answer: 850 cm = 8.5 m

3. Convert 5 kilometers (km) into meters (m)
What we know:
- 1 kilometer = 1,000 meters
Step 1: Multiply by 1,000: 5 × 1,000 = 5,000 m
Answer: 5 km = 5,000 m

4. Convert 2,000 millimeters (mm) into meters (m)
What we know:
- 1 meter = 1,000 millimeters
Step 1: Divide by 1,000: 2,000 ÷ 1,000 = 2 m
Answer: 2,000 mm = 2 m

5. Convert 3 kilograms (kg) into grams (g)
What we know:
- 1 kilogram = 1,000 grams
Step 1: Multiply by 1,000: 3 × 1,000 = 3,000 g
Answer: 3 kg = 3,000 g

6. Convert 7,500 grams (g) into kilograms (kg)
What we know:
- 1 kilogram = 1,000 grams
Step 1: Divide by 1,000: 7,500 ÷ 1,000 = 7.5 kg
Answer: 7,500 g = 7.5 kg

7. Convert 160 ounces (oz) into pounds (lb)
What we know:
- 1 pound = 16 ounces
Step 1: Divide by 16: 160 ÷ 16 = 10 lb
Answer: 160 oz = 10 lb

8. Convert 4 pounds (lb) into ounces (oz)
What we know:
- 1 pound = 16 ounces

Step 1: Multiply by 16: 4 × 16 = 64 oz
Answer: 4 lb = 64 oz

9. Convert 5 liters (L) into milliliters (mL)
What we know:
- 1 liter = 1,000 milliliters

Step 1: Multiply by 1,000: 5 × 1,000 = 5,000 mL
Answer: 5 L = 5,000 mL

10. Convert 2,500 milliliters (mL) into liters (L)
What we know:
- 1 liter = 1,000 milliliters

Step 1: Divide by 1,000: 2,500 ÷ 1,000 = 2.5 L
Answer: 2,500 mL = 2.5 L

11. Convert 1.5 meters (m) into centimeters (cm)
What we know:
- 1 meter = 100 centimeters

Step 1: Multiply by 100: 1.5 × 100 = 150 cm
Answer: 1.5 m = 150 cm

12. Convert 6 feet (ft) into inches (in)
What we know:
- 1 foot = 12 inches

Step 1: Multiply by 12: 6 × 12 = 72 in
Answer: 6 ft = 72 in

13. Convert 4 yards (yd) into feet (ft)
What we know:
- 1 yard = 3 feet

Step 1: Multiply by 3: 4 × 3 = 12 ft
Answer: 4 yd = 12 ft

14. Convert 3 miles (mi) into feet (ft)
What we know:
- 1 mile = 5,280 feet

Step 1: Multiply by 5,280: 3 × 5,280 = 15,840 ft
Answer: 3 mi = 15,840 ft

15. Convert 5 kilograms (kg) into grams (g)
What we know:
- 1 kilogram = 1,000 grams

Step 1: Multiply by 1,000: 5 × 1,000 = 5,000 g
Answer: 5 kg = 5,000 g

16. Convert 4,500 grams (g) into kilograms (kg)
What we know:
- 1 kilogram = 1,000 grams
Step 1: Divide by 1,000: 4,500 ÷ 1,000 = 4.5 kg
✅ **Answer: 4,500 g = 4.5 kg**

17. Convert 48 ounces (oz) into pounds (lb)
What we know:
- 1 pound = 16 ounces
Step 1: Divide by 16: 48 ÷ 16 = 3 lb
✅ **Answer: 48 oz = 3 lb**

18. Convert 3 pounds (lb) into ounces (oz)
What we know:
- 1 pound = 16 ounces
Step 1: Multiply by 16: 3 × 16 = 48 oz
✅ **Answer: 3 lb = 48 oz**

19. Convert 8 cups (c) into pints (pt)
What we know:
- 1 pint = 2 cups
Step 1: Divide by 2: 8 ÷ 2 = 4 pt
✅ **Answer: 8 cups = 4 pints**

20. Convert 3 quarts (qt) into cups (c)
What we know:
- 1 quart = 4 cups
Step 1: Multiply by 4: 3 × 4 = 12 cups
✅ **Answer: 3 qt = 12 cups**

21. Convert 10 millimeters (mm) into centimeters (cm)
What we know:
- 1 centimeter = 10 millimeters
Step 1: Divide by 10: 10 ÷ 10 = 1 cm
✅ **Answer: 10 mm = 1 cm**

22. Convert 5 meters (m) into centimeters (cm)
What we know:
- 1 meter = 100 centimeters
Step 1: Multiply by 100: 5 × 100 = 500 cm
✅ **Answer: 5 m = 500 cm**

23. Convert 3 kilometers (km) into meters (m)
What we know:
- 1 kilometer = 1,000 meters
Step 1: Multiply by 1,000: 3 × 1,000 = 3,000 m
✅ **Answer: 3 km = 3,000 m**

24. Convert 250 centimeters (cm) into meters (m)

What we know:

- 1 meter = 100 centimeters

Step 1: Divide by 100: 250 ÷ 100 = 2.5 m

✅ **Answer: 250 cm = 2.5 m**

25. Convert 2 liters (L) into milliliters (mL)

What we know:

- 1 liter = 1,000 milliliters

Step 1: Multiply by 1,000: 2 × 1,000 = 2,000 mL

✅ **Answer: 2 L = 2,000 mL**

26. Convert 1,500 milliliters (mL) into liters (L)

What we know:

- 1 liter = 1,000 milliliters

Step 1: Divide by 1,000: 1,500 ÷ 1,000 = 1.5 L

✅ **Answer: 1,500 mL = 1.5 L**

27. Convert 190 pounds (lbs) into kilograms (kg)

What we know:

- 1 pound ≈ 0.4536 kilograms

Step 1: Multiply by 0.4536: 190 × 0.4536 ≈ 86.18 kg

✅ **Answer: 190 lbs ≈ 86.18 kg**

28. Convert 60 minutes into hours

What we know:

- 1 hour = 60 minutes

Step 1: Divide by 60: 60 ÷ 60 = 1 hour

✅ **Answer: 60 minutes = 1 hour**

29. Convert 3 hours into minutes

What we know:

- 1 hour = 60 minutes

Step 1: Multiply by 60: 3 × 60 = 180 minutes

✅ **Answer: 3 hours = 180 minutes**

30. Convert 7 days into hours

What we know:

- 1 day = 24 hours

Step 1: Multiply by 24: 7 × 24 = 168 hours

✅ **Answer: 7 days = 168 hours**

Solutions: Representing and Interpreting Data: 1-5

1: Favorite Pets

Step 1: Identify the least popular pet:
Compare the numbers: Dogs = 10, Cats = 7, Fish = 5, Birds = 3. The smallest number is 3, which belongs to Birds.
✅ **Answer for (b): Birds are the least popular pet.**

Step 2: Calculate the difference between dogs and birds:
Dogs = 10, Birds = 3. Difference = 10 – 3 = 7.
✅ **Answer for (c): 7 more students like dogs than birds.**

2: Weekly Book Reading

Step 1: Identify the most books read:
Compare the numbers: Monday = 4, Tuesday = 6, Wednesday = 3, Thursday = 5, Friday = 7. The largest number is 7, which is on Friday.
✅ **Answer for (a): Friday had the most books read (7 books).**

Step 2: Identify the least books read:
The smallest number is 3, which is on Wednesday.

Step 3: Calculate the difference:
Most books = 7 (Friday), Least books = 3 (Wednesday). Difference = 7 – 3 = 4.
✅ **Answer for (b): The difference is 4 books.**

3: Fruit Sales

Step 1: Identify the fruit that sold the most:
Compare the numbers: Apples = 24, Bananas = 18, Grapes = 12, Oranges = 20. The largest number is 24, which belongs to Apples.
✅ **Answer for (a): Apples sold the most.**

Step 2: Calculate the difference between apples and grapes:
Apples = 24, Grapes = 12. Difference = 24 – 12 = 12.
✅ **Answer for (b): 12 more apples were sold than grapes.**

4: Sports Club Members

Step 1: Identify the second highest number of members:
Compare the numbers: Soccer = 15, Basketball = 10, Tennis = 8, Swimming = 12. The highest is Soccer (15). The second highest is Swimming (12).
✅ **Answer for (a): Swimming has the second highest number of members.**

Step 2: Calculate Tennis members after 3 more join:
Current Tennis members = 8. Add 3 more: 8 + 3 = 11.
✅ **Answer for (b): Tennis will have 11 members.**

5: Favorite Ice Cream Flavors

Step 1: Identify the most popular flavor:
Compare the numbers: Chocolate = 14, Vanilla = 9, Strawberry = 11, Mint = 6. The largest number is 14, which belongs to Chocolate.
✅ **Answer for (a): Chocolate is the most popular flavor.**

Step 2: Calculate the total votes:
Add all votes: 14 + 9 + 11 + 6 = 40.
✅ **Answer for (b): A total of 40 votes were cast.**

Solutions: Measurements and Data Post-Test

1. Composite Shape
A garden is shaped like an L.
- The long rectangle is 12 m by 4 m.
- The short rectangle is 4 m by 6 m.

(a) What is the total area of the garden?

Step 1: Find area of each rectangle: Long rectangle = $12 \times 4 = 48$ m²; Short rectangle = $4 \times 6 = 24$ m²
Step 2: Add areas: $48 + 24 = 72$ m²
✅ **Final Answer: 72 m²**

(b) What is the perimeter of the garden?

Step 1: Perimeter: Trace the outer boundary: $12 + 4 + 6 + 4 + 8 + 4 = 38$ m (8 m comes from $12 - 4$ due to overlap)
Step 2: Answer: Area = 72 m²; Perimeter = 38 m
✅ **Final Answer:** 38 m

2. Missing Side
A rectangular playground has an area of 96 m² and a perimeter of 40 m.
Find the length and width of the playground.

Step 1: Let length = L and width = W. Area: $L \times W = 96$; Perimeter: $2(L + W) = 40 \rightarrow L + W = 20$
Step 2: Use substitution: $W = 20 - L$
Step 3: Plug into area equation: $L \times (20 - L) = 96 \rightarrow 20L - L^2 = 96$
Step 4: Solve: $L^2 - 20L + 96 = 0$; Factors: $(L - 12)(L - 8) = 0 \rightarrow L = 12, W = 8$
Step 5: Answer: Length = 12 m; Width = 8 m
✅ **Final Answer:** 8 m

3. Angle Hunt in a Triangle
A triangle has three angles:
- Angle A = 65°
- Angle B = 90°
- Angle C = 25°

(a) Classify each angle as acute, right, or obtuse.

Step 1: Recall definitions: Acute < 90°; Right = 90°; Obtuse > 90°
Step 2: Classify each angle: A = 65° → Acute; B = 90° → Right; C = 25° → Acute
✅ **Final Answer:** Acute

(b) Which type of triangle is this based on its angles?

Step 1: Triangle type: It has one right angle → Right triangle
Step 2: Answer: Angles: A = Acute, B = Right, C = Acute; Triangle type = Right triangle
✅ **Final Answer:** Right triangle

4. Mixed Length & Time
Sofia is making a ribbon decoration. She has 3 pieces of ribbon:
- Piece A: 1.25 meters
- Piece B: 85 centimeters
- Piece C: 150 centimeters

She wants to join them into one long ribbon.

(a) What is the total length in meters?
Step 1: Convert all lengths to meters: Piece A = 1.25 m, Piece B = 85 cm = 0.85 m, Piece C = 150 cm = 1.50 m
Step 2: Add lengths: 1.25 + 0.85 + 1.50 = 3.60 meters
✅ **Final Answer:** 3.60 meters

(b) If she cuts the ribbon into 5 equal parts, how long is each part in centimeters?
Step 1: Divide into 5 equal parts: 360 ÷ 5 = 72 cm per part
Step 2: Answer: Total length = 3.6 meters; Each part = 72 cm
✅ **Final Answer:** 72 cm

5. Mixed Volume, Mass & Money

A juice shop sells orange juice in two sizes:

- Small bottle: 250 mL for $1.20
- Large bottle: 1.5 L for $5.40

A customer buys 3 small bottles and 2 large bottles.

(a) How many liters of juice does the customer buy?

Step 1: Convert small bottles to liters: 250 mL = 0.25 L; 3 small bottles = 0.25 × 3 = 0.75 L

Step 2: Large bottles: 2 large bottles = 1.5 × 2 = 3.0 L

Step 3: Total volume: 0.75 + 3.0 = 3.75 L

✅ **Final Answer:** 3.75 L

(b) What is the total cost?

Step 1: Calculate cost: Small bottles = $1.20 × 3 = $3.60; Large bottles = $5.40 × 2 = $10.80; Total = $14.40

✅ **Final Answer**$14.40

(c) If the customer pays $20, how much change will they get?

Step 1: Change from $20: $20.00 − $14.40 = $5.60

Step 2: Answer: Total juice = 3.75 L; Total cost = $14.40; Change = $5.60

✅ **Final Answer:** $5.60

I GOT THIS!
I GOT THIS!
I GOT THIS!

Certificate of Completion

This is proudly awarded to:

____

For successfully completing the
Math Practice Workbook: 4th Grade
Measurements and Data

This student has shown strong understanding of

- Solving word problems involving length, mass, volume, time & Money

Converting units within the same measurement system.

- Understanding area and perimeter
- Understanding types of angles (acute, obtuse, right)
- Representing and interpreting data using graphs and pictographs

Completed on: ________________________

Signed: ______________________________
(Parent/Teacher)

I GOT THIS!
I GOT THIS!
I GOT THIS!

Keep up the great work and continue your math journey! 🧠 ✨

Say it with me, "I Got This!"

Say it with me, "I Got This!"

Say it with me, "I Got This!"

Say it with me, "I Got This!"

Say it with me, "I Got This!"

Say it with me, "I Got This!"

Say it with me, "I Got This!"

Say it with me, "I Got This!"

Say it with me, "I Got This!"

Say it with me, "I Got This!"

Say it with me, "I Got This!"

Say it with me, "I Got This!"

Say it with me, "I Got This!"

Say it with me, "I Got This!"

Say it with me, "I Got This!"

Say it with me, "I Got This!"

Say it with me, "I Got This!"

Say it with me, "I Got This!"

Copyright & Disclaimer

Educational Purpose Only

This workbook was carefully crafted to help 4th grade students build confidence and develop strong math skills in a fun and approachable way. While every effort has been made to ensure accuracy, Chelse's Creations makes no guarantees regarding outcomes and is not responsible for errors or omissions.

Thank You

Thank you for supporting creative educators and resources made with care. By using this workbook, you are encouraging learning, imagination, and growth. A portion of all proceeds goes back into the community through our volunteer organization, Helping from the Heart Foundation. Thank you for making a difference!

For more learning resources, visit: **Chelse's Creations Amazon Store**

About the Author

Chelse Burton is an accomplished mom, nurse, author, and artist who is passionate about education and helping others. Her diverse talents and dedication shine through everything she creates—from inspiring novels and educational children's books to beautiful artwork and even her own line of homemade seasonings!

With a background as a trauma nurse, educator, and case manager, Chelse has seen both the strength and struggles of people in her community. These experiences motivate her to share messages of hope, learning, and creativity through her work.

Chelse's passion for education shines brightest in her children's books and educational series, "I Got This!". Through engaging stories and fun workbooks, she encourages young learners to explore, think deeply, and believe in themselves. Her mission is to inspire curiosity and confidence in every child—helping them see that learning can be both exciting and empowering!

Chelse and her husband Victor, a proud Marine veteran, are the founders of Helping from the Heart Foundation, a volunteer organization dedicated to serving the community with the same dedication, integrity, and leadership that defines the Marine Corps. The foundation is committed to making a lasting impact through volunteer efforts that support those in need—especially fellow veterans, first responders, and underserved communities.

A portion of all proceeds made from Chelse's Creations, including this workbook, goes back into the community through the volunteer organization, Helping from the Heart Foundation. *Thank you* for making a difference!

Made in the USA
Monee, IL
07 November 2026

68031434R00057